The Theory of Parallels

N. Lobatchewsky

GEOMETRICAL RESEARCHES

ON

THE THEORY OF PARALLELS

BY

NICHOLAS LOBACHEVSKI

Imperial Russian Real Councillor of State and Regular Professor
of Mathematics in the University of Kasan

BERLIN, 1840

Translated from the Original

BY

GEORGE BRUCE HALSTED

A. M., Ph. D., Ex-Fellow of Princeton and
Johns-Hopkins University

NEW EDITION

Watchmaker Publishing

1891

Planographed by John S. Swift Co., Inc., Chicago, St. Louis, New York, Cincinnati

TRANSLATOR'S PREFACE.

Lobachevski was the first man ever to publish a non-Euclidean geometry.

Of the immortal essay now first appearing in English Gauss said, "The author has treated the matter with a master-hand and in the true geometer's spirit. I think I ought to call your attention to this book, whose perusal can not fail to give you the most vivid pleasure."

Clifford says, "It is quite simple, merely Euclid without the vicious assumption, but the way things come out of one another is quite lovely." * * * "What Vesalius was to Galen, what Copernicus was to Ptolemy, that was Lobachevski to Euclid."

Says Sylvester, "In Quaternions the example has been given of Algebra released from the yoke of the commutative principle of multiplication—an emancipation somewhat akin to Lobachevski's of Geometry from Euclid's noted empirical axiom."

Cayley says, "It is well known that Euclid's twelfth axiom, even in Playfair's form of it, has been considered as needing demonstration; and that Lobachevski constructed a perfectly consistent theory, wherein this axiom was assumed not to hold good, or say a system of non-Euclidean plane geometry. There is a like system of non-Euclidean solid geometry."

GEORGE BRUCE HALSTED.

2407 San Marcos Street,
 Austin, Texas.
May 1, 1891.

TRANSLATOR'S INTRODUCTION.

"Prove all things, hold fast that which is good," does not mean demonstrate everything. From nothing assumed, nothing can be proved. "Geometry without axioms," was a book which went through several editions, and still has historical value. But now a volume with such a title would, without opening it, be set down as simply the work of a paradoxer.

The set of axioms far the most influential in the intellectual history of the world was put together in Egypt; but really it owed nothing to the Egyptian race, drew nothing from the boasted lore of Egypt's priests.

The Papyrus of the Rhind, belonging to the British Museum, but given to the world by the erudition of a German Egyptologist, Eisenlohr, and a German historian of mathematics, Cantor, gives us more knowledge of the state of mathematics in ancient Egypt than all else previously accessible to the modern world. Its whole testimony confirms with overwhelming force the position that Geometry as a science, strict and self-conscious deductive reasoning, was created by the subtle intellect of the same race whose bloom in art still overawes us in the Venus of Milo, the Apollo Belvidere, the Laocoon.

In a geometry occur the most noted set of axioms, the geometry of Euclid, a pure Greek, professor at the University of Alexandria.

Not only at its very birth did this typical product of the Greek genius assume sway as ruler in the pure sciences, not only does its first efflorescence carry us through the splendid days of Theon and Hypatia, but unlike the latter, fanatics can not murder it; that dismal flood, the dark ages, can not drown it. Like the phœnix of its native Egypt, it rises with the new birth of culture. An Anglo-Saxon, Adelard of Bath, finds it clothed in Arabic vestments in the land of the Alhambra. Then clothed in Latin, it and the new-born printing press confer honor on each other. Finally back again in its original Greek, it is published first in queenly Basel, then in stately Oxford. The latest edition in Greek is from Leipsic's learned presses.

[5]

How the first translation into our cut-and-thrust, survival-of-the-fittest English was made from the Greek and Latin by Henricus Billingsly, Lord Mayor of London, and published with a preface by John Dee the Magician, may be studied in the Library of our own Princeton, where they have, by some strange chance, Billingsly's own copy of the Arabic-Latin version of Campanus bound with the Editio Princeps in Greek and enriched with his autograph emendations. Even to-day in the vast system of examinations set by Cambridge, Oxford, and the British government, no proof will be accepted which infringes Euclid's order, a sequence founded upon his set of axioms.

The American ideal is success. In twenty years the American maker expects to be improved upon, superseded. The Greek ideal was perfection. The Greek Epic and Lyric poets, the Greek sculptors, remain unmatched. The axioms of the Greek geometer remained unquestioned for twenty centuries.

How and where doubt came to look toward them is of no ordinary interest, for this doubt was epoch-making in the history of mind.

Among Euclid's axioms was one differing from the others in prolixity, whose place fluctuates in the manuscripts, and which is not used in Euclid's first twenty-seven propositions. Moreover it is only then brought in to prove the inverse of one of these already demonstrated.

All this suggested, at Europe's renaissance, not a doubt of the axiom, but the possibility of getting along without it, of deducing it from the other axioms and the twenty-seven propositions already proved. Euclid demonstrates things more axiomatic by far. He proves what every dog knows, that any two sides of a triangle are together greater than the third. Yet when he has perfectly proved that lines making with a transversal equal alternate angles are parallel, in order to prove the inverse, that parallels cut by a transversal make equal alternate angles, he brings in the unwieldly postulate or axiom:

"If a straight line meet two straight lines, so as to make the two interior angles on the same side of it taken together less than two right angles, these straight lines, being continually produced, shall at length meet on that side on which are the angles which are less than two right angles."

Do you wonder that succeeding geometers wished by demonstration to push this unwieldly thing from the set of fundamental axioms.

Numerous and desperate were the attempts to deduce it from reasonings about the nature of the straight line and plane angle. In the "Encyclopœdie der Wissenschaften und Kunste; Von Ersch und Gruber;" Leipzig, 1838; under "Parallel," Sohncke says that in mathematics there is nothing over which so much has been spoken, written, and striven, as over the theory of parallels, and all, so far (up to his time), without reaching a definite result and decision.

Some acknowledged defeat by taking a new definition of parallels, as for example the stupid one, "Parallel lines are everywhere equally distant," still given on page 33 of Schuyler's Geometry, which that author, like many of his unfortunate prototypes, then attempts to identify with Euclid's definition by pseudo-reasoning which tacitly assumes Euclid's postulate, e. g. he says p. 35: "For, if not parallel, they are not everywhere equally distant; and since they lie in the same plane; must approach when produced one way or the other; and since straight lines continue in the same direction, must continue to approach if produced farther, and if sufficiently produced, must meet." This is nothing but Euclid's assumption, diseased and contaminated by the introduction of the indefinite term "direction."

How much better to have followed the third class of his predecessors who honestly assume a new axiom differing from Euclid's in form if not in essence. Of these the best is that called Playfair's; "Two lines which intersect can not both be parallel to the same line."

The German article mentioned is followed by a carefully prepared list of ninety-two authors on the subject. In English an account of like attempts was given by Perronet Thompson, Cambridge, 1833, and is brought up to date in the charming volume, "Euclid and his Modern Rivals," by C. L. Dodgson, late Mathematical Lecturer of Christ Church, Oxford, the Lewis Carroll, author of Alice in Wonderland.

All this shows how ready the world was for the extraordinary flaming-forth of genius from different parts of the world which was at once to overturn, explain, and remake not only all this subject but as consequence all philosophy, all ken-lore. As was the case with the discovery of the Conservation of Energy, the independent irruptions of genius, whether in Russia, Hungary, Germany, or even in Canada gave everywhere the same results.

At first these results were not fully understood even by the brightest intellects. Thirty years after the publication of the book he mentions, we see the brilliant Clifford writing from Trinity College, Cambridge, April 2, 1870, "Several new ideas have come to me lately: First I have procured Lobachevski, 'Études Géométriques sur la Théorie des Parallels' – – – a small tract of which Gauss, therein quoted, says: L'auteur a traité la matière en main de maître et avec le véritable esprit géométrique. Je crois devoir appeler votre attention sur ce livre, dont la lecture ne peut manquer de vous causer le plus vif plaisir.'" Then says Clifford: "It is quite simple, merely Euclid without the vicious assumption, but the way the things come out of one another is quite lovely."

The first axiom doubted is called a "vicious assumption," soon no man sees more clearly than Clifford that all are assumptions and none vicious. He had been reading the French translation by Hoüel, published in 1866, of a little book of 61 pages published in 1840 in Berlin under the title Geometrische Untersuchungen zur Theorie der Parallellinien by Nicolas Lobachevski (1793-1856), the first public expression of whose discoveries, however, dates back to a discourse at Kasan on February 12, 1826.

Under this commonplace title who would have suspected the discovery of a new space in which to hold our universe and ourselves.

A new kind of universal space; the idea is a hard one. To name it, all the space in which we think the world and stars live and move and have their being was ceded to Euclid as his by right of pre-emption, description, and occupancy; then the new space and its quick-following fellows could be called Non-Euclidean.

Gauss in a letter to Schumacher, dated Nov. 28, 1846, mentions that as far back as 1792 he had started on this path to a new universe. Again he says: "La géométrie non-euclidienne ne renferme en elle rien de contradictoire, quoique, à prèmiere vue, beaucoup de ses résultats aien l'air de paradoxes. Ces contradictions apparents doivent être regardées comme l'effet d'une illusion, due à l'habitude que nous avons prise de bonne heure de considérer la géométrie euclidienne comme rigoureuse."

But here we see in the last word the same imperfection of view as in Clifford's letter. The perception has not yet come that though the non-Euclidean geometry is rigorous, Euclid is not one whit less so.

A former friend of Gauss at Gœttingen was the Hungarian Wolfgang Bolyai. His principal work, published by subscription, has the following title:

Tentamen Juventutem studiosam in elementa Matheseos purae, elementaris ac sublimioris, methodo intuitiva, evidentiaque huic propria, introducendi. Tomus Primus, 1832; Secundus, 1833. 8vo. Maros-Vásárhelyini.

In the first volume with special numbering, appeared the celebrated Appendix of his son John Bolyai with the following title:

APPENDIX.

Scientiam spatii *absolute veram* exhibens: *a veritate aut falsitate Axiomatis XI Euclidei (a priori haud unquam decidenda) independentem.* Auctore Johanne Bolyai de eadem, Geometrarum in Exercitu Caesareo Regio Austriaco Castrensium Capitaneo. (26 pages of text).

This marvellous Appendix has been translated into French, Italian, English and German.

In the title of Wolfgang Bolyai's last work, the only one he composed in German (88 pages of text, 1851), occurs the following:

"und da die Frage, *ob zwey von der dritten geschnittene Geraden, wenn die summe der inneren Winkel nicht=2R, sich schneiden oder nicht?* niemand auf der Erde ohne ein Axiom (wie *Euclid* das XI) aufzustellen, beantworten wird; die davon unabhængige Geometrie abzusondern; und eine auf die *Ja*-Antwort, andere auf das *Nein* so zu bauen, dass die Formeln der letzen, auf ein Wink auch in der ersten gültig seyen."

The author mentions Lobachevski's Geometrische Untersuchungen, Berlin, 1840, and compares it with the work of his son John Bolyai, "au sujet duquel il dit: 'Quelques exemplaires de l'ouvrage publié ici ont été envoyés à cette époque à Vienne, à Berlin, à Gœttingue. . . De Goettingue le géant mathématique, [Gauss] qui du sommet des hauteurs embrasse du même regard les astres et la profondeur des abîmes, a écrit qu'il était ravi de voir exécuté le travail qu'il avait commencé pour le laisser après lui dans ses papiers.'"

In fact this first of the Non-Euclidean geometries accepts all of Euclid's axioms but the last, which it flatly denies and replaces by its contradictory, that the sum of the interior angles made on the same side of a transversal by two straight lines may be less than a straight angle without the lines meeting. A perfectly consistent and elegant geometry then follows, in which the sum of the angles of a triangle is always less than a straight angle, and not every triangle has its vertices concyclic.

THEORY OF PARALLELS.

In geometry I find certain imperfections which I hold to be the reason why this science, apart from transition into analytics, can as yet make no advance from that state in which it has come to us from Euclid.

As belonging to these imperfections, I consider the obscurity in the fundamental concepts of the geometrical magnitudes and in the manner and method of representing the measuring of these magnitudes, and finally the momentous gap in the theory of parallels, to fill which all efforts of mathematicians have been so far in vain.

For this theory Legendre's endeavors have done nothing, since he was forced to leave the only rigid way to turn into a side path and take refuge in auxiliary theorems which he illogically strove to exhibit as necessary axioms. My first essay on the foundations of geometry I published in the Kasan *Messenger* for the year 1829. In the hope of having satisfied all requirements, I undertook hereupon a treatment of the whole of this science, and published my work in separate parts in the *"Gelehrten Schriften der Universitæt Kasan"* for the years 1836, 1837, 1838, under the title "New Elements of Geometry, with a complete Theory of Parallels." The extent of this work perhaps hindered my countrymen from following such a subject, which since Legendre had lost its interest. Yet I am of the opinion that the Theory of Parallels should not lose its claim to the attention of geometers, and therefore I aim to give here the substance of my investigations, remarking beforehand that contrary to the opinion of Legendre, all other imperfections — for example, the definition of a straight line — show themselves foreign here and without any real influence on the theory of parallels.

In order not to fatigue my reader with the multitude of those theorems whose proofs present no difficulties, I prefix here only those of which a knowledge is necessary for what follows.

1. A straight line fits upon itself in all its positions. By this I mean that during the revolution of the surface containing it the straight line does not change its place if it goes through two unmoving points in the surface: (*i. e.*, if we turn the surface containing it about two points of the line, the line does not move.)

[11]

2. Two straight lines can not intersect in two points.

3. A straight line sufficiently produced both ways must go out beyond all bounds, and in such way cuts a bounded plain into two parts.

4. Two straight lines perpendicular to a third never intersect, how far soever they be produced.

5. A straight line always cuts another in going from one side of it over to the other side: (*i. e.*, one straight line must cut another if it has points on both sides of it.)

6. Vertical angles, where the sides of one are productions of the sides of the other, are equal. This holds of plane rectilineal angles among themselves, as also of plane surface angles: (*i. e.*, dihedral angles.)

7. Two straight lines can not intersect, if a third cuts them at the same angle.

8. In a rectilineal triangle equal sides lie opposite equal angles, and inversely.

9. In a rectilineal triangle, a greater side lies opposite a greater angle. In a right-angled triangle the hypothenuse is greater than either of the other sides, and the two angles adjacent to it are acute.

10. Rectilineal triangles are congruent if they have a side and two angles equal, or two sides and the included angle equal, or two sides and the angle opposite the greater equal, or three sides equal.

11. A straight line which stands at right angles upon two other straight lines not in one plane with it is perpendicular to all straight lines drawn through the common intersection point in the plane of those two.

12. The intersection of a sphere with a plane is a circle.

13. A straight line at right angles to the intersection of two perpendicular planes, and in one, is perpendicular to the other.

14. In a spherical triangle equal sides lie opposite equal angles, and inversely.

15. Spherical triangles are congruent (or symmetrical) if they have two sides and the included angle equal, or a side and the adjacent angles equal.

From here follow the other theorems with their explanations and proofs.

16. All straight lines which in a plane go out from a point can, with reference to a given straight line in the same plane, be divided into two classes — into *cutting* and *not-cutting*.

The *boundary lines* of the one and the other class of those lines will be called *parallel to the given line*.

From the point A (Fig. 1) let fall upon the line BC the perpendicular AD, to which again draw the perpendicular AE.

In the right angle EAD either will all straight lines which go out from the point A meet the line DC, as for example AF, or some of them, like the perpendicular AE, will not meet the line DC. In the uncertainty whether the perpendicular AE is the only line which does not meet DC, we will assume it may be possible that there are still other lines, for example AG, which do not cut DC, how far soever they may be prolonged. In passing over from the cutting lines, as AF, to the not-cutting lines, as AG, we must come upon a line AH, parallel to DC, a boundary line, upon one side of which all lines AG are such as do not meet the line DC, while upon the other side every straight line AF cuts the line DC.

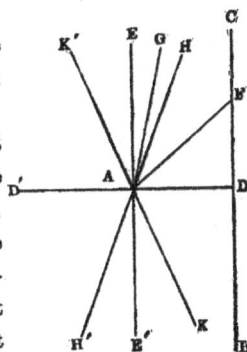

Fig. 1.

The angle HAD between the parallel HA and the perpendicular AD is called the parallel angle (angle of parallelism), which we will here designate by Π (p) for AD = p.

If Π (p) is a right angle, so will the prolongation AE' of the perpendicular AE likewise be parallel to the prolongation DB of the line DC, in addition to which we remark that in regard to the four right angles, which are made at the point A by the perpendiculars AE and AD, and their prolongations AE' and AD', every straight line which goes out from the point A, either itself or at least its prolongation, lies in one of the two right angles which are turned toward BC, so that except the parallel EE' all others, if they are sufficiently produced both ways, must intersect the line BC.

If Π (p) < $\frac{1}{2}$ π, then upon the other side of AD, making the same angle DAK = Π (p) will lie also a line AK, parallel to the prolongation DB of the line DC, so that under this assumption we must also make a distinction of *sides in parallelism*.

All remaining lines or their prolongations within the two right angles turned toward BC pertain to those that intersect, if they lie within the angle $HAK = 2 \, \varPi \, (p)$ between the parallels; they pertain on the other hand to the non-intersecting AG, if they lie upon the other sides of the parallels AH and AK, in the opening of the two angles $EAH = \frac{1}{2} \pi - \varPi \, (p)$, $E'AK = \frac{1}{2} \pi - \varPi \, (p)$, between the parallels and EE' the perpendicular to AD. Upon the other side of the perpendicular EE' will in like manner the prolongations AH' and AK' of the parallels AH and AK likewise be parallel to BC; the remaining lines pertain, if in the angle K'AH', to the intersecting, but if in the angles K'AE, H'AE' to the non-intersecting.

In accordance with this, for the assumption $\varPi(p) = \frac{1}{2} \pi$. the lines can be only intersecting or parallel; but if we assume that $\varPi(p) < \frac{1}{2} \pi$, then we must allow two parallels, one on the one and one on the other side; in addition we must distinguish the remaining lines into non-intersecting and intersecting.

For both assumptions it serves as the mark of parallelism that the line becomes intersecting for the smallest deviation toward the side where lies the parallel, so that if AH is parallel to DC, every line AF cuts DC, how small soever the angle HAF may be.

17. *A straight line maintains the characteristic of parallelism at all its points.*

Given AB (Fig. 2) parallel to CD, to which latter AC is **perpendic**

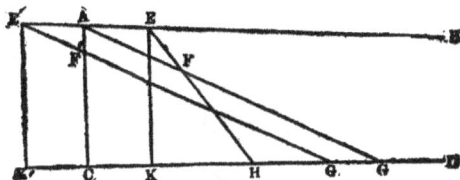

FIG. 2.

ular.   We will consider two points taken at random on the line AB and its production beyond the perpendicular.

Let the point E lie on that side of the perpendicular on which AB is looked upon as parallel to CD.

Let fall from the point E a perpendicular EK on CD and so draw EF that it falls within the angle BEK.

Connect the points A and F by a straight line, whose production then (by Theorem 16) must cut CD somewhere in G. Thus we get a triangle ACG, into which the line EF goes; now since this latter, from the construction, can not cut AC, and can not cut AG or EK a second time (Theorem 2), therefore it must meet CD somewhere at H (Theorem 3).

Now let E' be a point on the production of AB and E'K' perpendicular to the production of the line CD; draw the line E'F' making so small an angle AE'F' that it cuts AC somewhere in F'; making the same angle with AB, draw also from A the line AF, whose production will cut CD in G (Theorem 16.)

Thus we get a triangle AGC, into which goes the production of the line E'F'; since now this line can not cut AC a second time, and also can not cut AG, since the angle BAG = BE'G', (Theorem 7), therefore must it meet CD somewhere in G'.

Therefore from whatever points E and E' the lines EF and E'F' go out, and however little they may diverge from the line AB, yet will they always cut CD, to which AB is parallel.

18. *Two lines are always mutually parallel.*

Let AC be a perpendicular on CD, to which AB is parallel
if we draw from C the line
CE making any acute angle
ECD with CD, and let fall
from A the perpendicular AF
upon CE, we obtain a right-
angled triangle ACF, in which
AC, being the hypothenuse,
is greater than the side AF
(Theorem 9.)

Make AG = AF, and slide

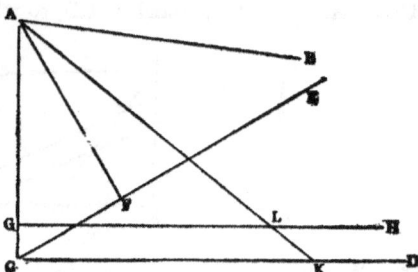

Fig. 3.

the figure EFAB until AF coincides with AG, when AB and FE will
take the position AK and GH, such that the angle BAK = FAC, con
sequently AK must cut the line DC somewhere in K (Theorem 16), thus
forming a triangle AKC, on one side of which the perpendicular GH
intersects the line AK in L (Theorem 3), and thus determines the distance AL of the intersection point of the lines AB and CE on the line
AB from the point A.

Hence it follows that CE will always intersect AB, how small soever
may be the angle ECD, consequently CD is parallel to AB (Theorem 16.)

19. *In a rectilineal triangle the sum of the three angles can not be greater
than two right angles.*

Suppose in the triangle ABC (Fig. 4) the sum of the three angles is
equal to $\pi + a$; then choose in case
of the inequality of the sides the
smallest BC, halve it in D, draw
from A through D the line AD
and make the prolongation of it,
DE, equal to AD, then join the
point E to the point C by the

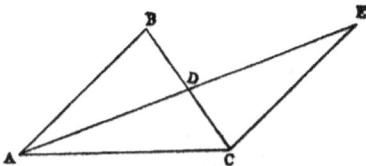

Fig. 4.

straight line EC. In the congruent triangles ADB and CDE, the angle
ABD = DCE, and BAD = DEC (Theorems 6 and 10); whence follows
that also in the triangle ACE the sum of the three angles must be equal
to $\pi + a$; but also the smallest angle BAC (Theorem 9) of the triangle
ABC in passing over into the new triangle ACE has been cut up into
the two parts EAC and AEC. Continuing this process, continually

halving the side opposite the smallest angle, we must finally attain to a triangle in which the sum of the three angles is $\pi + a$, but wherein are two angles, each of which in absolute magnitude is less than $\frac{1}{2}a$; since now, however, the third angle can not be greater than π, so must a be either null or negative.

20. *If in any rectilineal triangle the sum of the three angles is equal to two right angles, so is this also the case for every other triangle.*

If in the rectilineal triangle ABC (Fig. 5) the sum of the three angles $= \pi$, then must at least two of its angles, A and C, be acute. Let fall from the vertex of the third angle B upon the opposite side AC the perpendicular p. This will cut the tri-angle into two right-angled triangles, in each

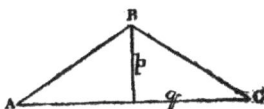

Fig. 5.

of which the sum of the three angles must also be π, since it can not in either be greater than π, and in their combination not less than π.

So we obtain a right-angled triangle with the perpendicular sides p and q, and from this a quadrilateral whose opposite sides are equal and whose adjacent sides p and q are at right angles (Fig. 6.)

By repetition of this quadrilateral we can make another with sides np and q, and finally a quadrilateral ABCD with sides at right angles to each other, such that AB = np, AD = mq, DC = np, BC = mq, where

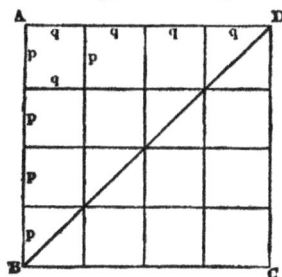

Fig. 6.

m and n are any whole numbers. Such a quadrilateral is divided by the diagonal DB into two congruent right-angled triangles, BAD and BCD, in each of which the sum of the three angles $= \pi$.

The numbers n and m can be taken sufficiently great for the right-angled triangle ABC (Fig. 7) whose perpendicular sides AB = np, BC = mq, to enclose within itself another given (right-angled) triangle BDE as soon as the right-angles fit each other.

2 — par.

Drawing the line DC, we obtain right-angled triangles of which every successive two have a side in common.

The triangle ABC is formed by the union of the two triangles ACD and DCB, in neither of which can the sum of the angles be greater than π; consequently it must be equal to π, in order that the sum in the compound triangle may be equal to π.

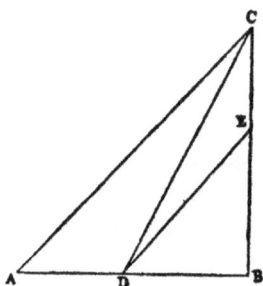

FIG. 7.

In the same way the triangle BDC consists of the two triangles DEC and DBE, consequently must in DBE the sum of the three angles be equal to π, and in general this must be true for every triangle, since each can be cut into two right-angled triangles.

From this it follows that only two hypotheses are allowable: Either is the sum of the three angles in all rectilineal triangles equal to π, or this sum is in all less than π.

21. *From a given point we can always draw a straight line that shall make with a given straight line an angle as small as we choose.*

Let fall from the given point A (Fig. 8) upon the given line BC the

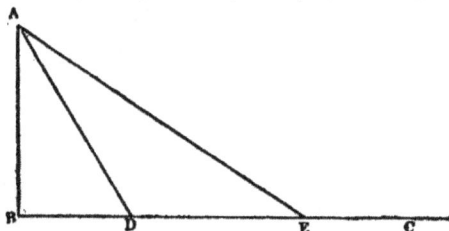

FIG. 8.

perpendicular AB; take upon BC at random the point D; draw the line AD; make DE = AD, and draw AE.

In the right-angled triangle ABD let the angle ADB = a; then must in the isosceles triangle ADE the angle AED be either $\frac{1}{2}a$ or less (Theorems 8 and 20). Continuing thus we finally attain to such an angle, AEB, as is less than any given angle.

22. *If two perpendiculars to the same straight line are parallel to each other, then the sum of the three angles in a rectilineal triangle is equal to two right angles.*

Let the lines AB and CD (Fig. 9) be parallel to each other and perpendicular to AC.

Draw from A the lines AE and AF to the points E and F, which are taken on the line CD at any distances FC > EC from the point C.

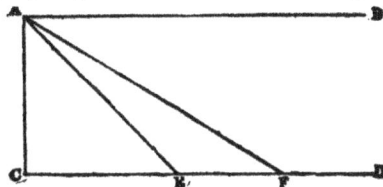

FIG. 9.

Suppose in the right-angled triangle ACE the sum of the three angles is equal to $\pi - a$, in the triangle AEF equal to $\pi - \beta$, then must it in triangle ACF equal $\pi - a - \beta$, where a and β can not be negative.

Further, let the angle BAF = a, AFC = b, so is $a + \beta = a - b$; now by revolving the line AF away from the perpendicular AC we can make the angle a between AF and the parallel AB as small as we choose; so also can we lessen the angle b, consequently the two angles a and β can have no other magnitude than $a = 0$ and $\beta = 0$.

It follows that in all rectilineal triangles the sum of the three angles is either π and at the same time also the parallel angle Π (p) = $\frac{1}{2} \pi$ for every line p, or for all triangles this sum is $< \pi$ and at the same time also Π(p) $< \frac{1}{2} \pi$.

The first assumption serves as *foundation for the ordinary geometry and plane trigonometry.*

The second assumption can likewise be admitted without leading to any contradiction in the results, and founds a new geometric science, to which I have given the name *Imaginary Geometry*, and which I intend here to expound as far as the development of the equations between the sides and angles of the rectilineal and spherical triangle.

23. *For every given angle a t h e r e i s a line p such that Π (p) = a.*

Let AB and AC (Fig. 10) be two straight lines which at the intersection point A make the acute angle a; take at random on AB a point

B'; from this point drop B'A' at right angles to AC; make A'A" = AA'; erect at A" the perpendicular A"B'; and so continue until a per-

Fɪɢ. 10.

pendicular CD is attained, which no longer intersects AB.   This must of necessity happen, for if in the triangle AA'B' the sum of all three angles is equal to $\pi - a$, then in the triangle AB'A" it equals $\pi - 2a$, in triangle AA"B" less than $\pi - 2a$ (Theorem 20), and so forth, until it finally becomes negative and thereby shows the impossibility of con-structing the triangle.

The perpendicular CD may be the very one nearer than which to the point A all others cut AB; at least in the passing over from those that cut to those not cutting such a perpendicular FG must exist.

Draw now from the point F the line FH, which makes with FG the acute angle HFG, on that side where lies the point A. From any point H of the line FH let fall upon AC the perpendicular HK, whose pro-longation consequently must cut AB somewhere in B, and so makes a triangle AKB, into which the prolongation of the line FH enters, and therefore must meet the hypothenuse AB somewhere in M.   Since the angle GFH is arbitrary and can be taken as small as we wish, therefore FG is parallel to AB and AF = p.   (Theorems 16 and 18.)

One easily sees that with the lessening of p the angle a increases, while, for p = 0, it approaches the value $\frac{1}{2}\pi$; with the growth of p the angle a decreases, while it continually approaches zero for p = ∞.

Since we are wholly at liberty to choose what angle we will understand by the symbol Π (p) when the line p is expressed by a negative number, so we will assume

$$\Pi(p)+\Pi(-p)=\pi,$$

an equation which shall hold for all values of p, positive as well as negative, and for $p=0$.

24. *The farther parallel lines are prolonged on the side of their parallelism, the more they approach one another.*

If to the line AB (Fig. 11) two perpendiculars $AC=BD$ are erected and their end-points C and D joined by a straight line, then will the quadrilateral CABD have two right angles at A and B, but two acute angles at C and D (Theorem 22) which are equal to one another, as we can easily see by thinking the quadrilateral super-imposed upon itself so that the line BD falls upon AC and AC upon BD.

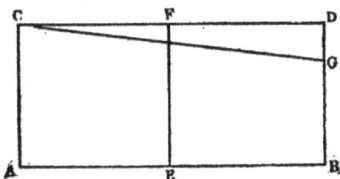

Fig. 11.

Halve AB and erect at the mid-point E the line EF perpendicular to AB. This line must also be perpendicular to CE, since the quadrilaterals CAEF and FDBE fit one another if we so place one on the other that the line EF remains in the same position. Hence the line CD can not be parallel to AB, but the parallel to AB for the point C, namely CG, must incline toward AB (Theorem 16) and cut from the perpendicular BD a part $BG < CA$.

Since C is a random point in the line CG, it follows that CG itself nears AB the more the farther it is prolonged.

25. *Two straight lines which are parallel to a third are also parallel to each other.*

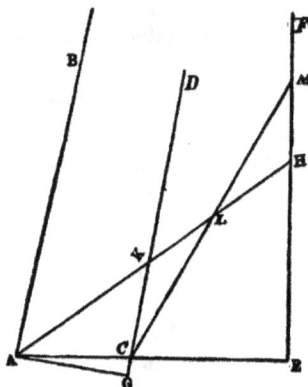

FIG. 12.

We will first assume that the three lines AB, CD, EF (Fig. 12) lie in one plane.   If two of them in order, AB and CD, are parallel to the outmost one, EF, so are AB and CD parallel to each other.   In order to prove this, let fall from any point A of the outer line AB upon the other outer line FE, the perpendicular AE, which will cut the middle line CD in some point C (Theorem 3), at an angle DCE $< \frac{1}{2}\pi$ on the side toward EF, the parallel to CD (Theorem 22).

A perpendicular AG let fall upon CD from the same point, A, must fall within the opening of the acute angle ACG (Theorem 9); every other line AH from A drawn within the angle BAC must cut EF, the parallel to AB, somewhere in H, how small soever the angle BAH may be; consequently will CD in the triangle AEH cut the line AH somewhere in K, since it is impossible that it should meet EF.   If AH from the point A went out within the angle CAG, then must it cut the prolongation of CD between the points C and G in the triangle CAG. Hence follows that AB and CD are parallel (Theorems 16 and 18).

Were both the outer lines AB and EF assumed parallel to the middle line CD, so would every line AK from the point A, drawn within the angle BAE, cut the line CD somewhere in the point K, how small soever the angle BAK might be.

Upon the prolongation of AK take at random a point L and join it with C by the line CL, which must cut EF somewhere in M, thus making a triangle MCE.

The prolongation of the line AL within the triangle MCE can cut neither AC nor CM a second time, consequently it must meet EF somewhere in H; therefore AB and EF are mutually parallel.

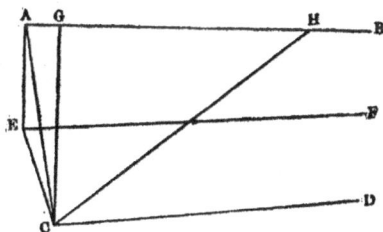

FIG. 13.

Now let the parallels AB and CD (Fig. 13) lie in two planes whose intersection line is EF. From a random point E of this latter let fall a perpendicular EA upon one of the two parallels, e. g., upon AB, then from A, the foot of the perpendicular EA, let fall a new perpendicular AC upon the other parallel CD and join the end-points E and C of the two perpendiculars by the line EC. The angle BAC must be acute (Theorem 22), consequently a perpendicular CG from C let fall upon AB meets it in the point G upon that side of CA on which the lines AB and CD are considered as parallel.

Every line EH [in the plane FEAB], however little it diverges from EF, pertains with the line EC to a plane which must cut the plane of the two parallels AB and CD along some line CH. This latter line cuts AB somewhere, and in fact in the very point H which is common to all three planes, through which necessarily also the line EH goes; consequently EF is parallel to AB.

In the same way we may show the parallelism of EF and CD.

Therefore the hypothesis that a line EF is parallel to one of two other parallels, AB and CD, is the same as considering EF as the intersection of two planes in which two parallels, AB, CD, lie.

Consequently two lines are parallel to one another if they are parallel to a third line, though the three be not co-planar.

The last theorem can be thus expressed:

Three planes intersect in lines which are all parallel to each other if the parallelism of two is pre-supposed.

26. *Triangles standing opposite to one another on the sphere are equivalent in surface.*

By opposite triangles we here understand such as are made on both sides of the center by the intersections of the sphere with planes; in such triangles, therefore, the sides and angles are in contrary order.

In the opposite triangles ABC and A′B′C′ (Fig. 14, where one of them must be looked upon as represented turned about), we have the sides AB = A′B′, BC = B′C′, CA = C′A′, and the corresponding angles

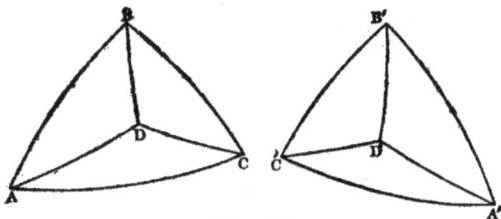

Fɪɢ. 14.

at the points A, B, C are likewise equal to those in the other triangle at th points A′, B′, C′.

Through the three points A, B, C, suppose a plane passed, and upon it from the center of the sphere a perpendicular dropped whose prolongations both ways cut both opposite triangles in the points D and D′ of the sphere. The distances of the first D from the points ABC, in arcs of great circles on the sphere, must be equal (Theorem 12) as well to each other as also to the distances D′A′, D′B′, D′C′, on the other triangle (Theorem 6), consequently the isosceles triangles about the points D and D′ in the two spherical triangles ABC and A′B′C′ are congruent.

In order to judge of the equivalence of any two surfaces in general, I take the following theorem as fundamental:

Two surfaces are equivalent when they arise from the mating or separating of equal parts.

27. *A three-sided solid angle equals the half sum of the surface angles less a right-angle.*

In the spherical triangle ABC (Fig. 15), where each side $< \pi$, designate the angles by A, B, C; prolong the side AB so that a whole circle ABA′B′A is produced; this divides the sphere into two equal parts.

In that half in which is the triangle ABC, prolong now the other two sides through their common intersection point C until they meet the circle in A' and B'.

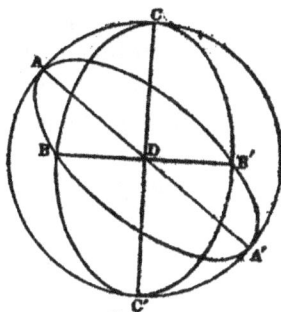

FIG. 15.

In this way the hemisphere is divided into four triangles, ABC, ACB', B'CA', A'CB, whose size may be designated by P, X, Y, Z. It is evident that here $P + X = B$, $P + Z = A$.

The size of the spherical triangle Y equals that of the opposite triangle ABC', having a side AB in common with the triangle P, and whose third angle C' lies at the end-point of the diameter of the sphere which goes from C through the center D of the sphere (Theorem 26). Hence it follows that

$P + Y = C$, and since $P + X + Y + Z = \pi$, therefore we have also

$$P = \tfrac{1}{2}(A + B + C - \pi).$$

We may attain to the same conclusion in another way, based solely upon the theorem about the equivalence of surfaces given above. (Theorem 26.)

In the spherical triangle ABC (Fig. 16), halve the sides AB and BC, and through the mid-points D and E draw a great circle; upon this let fall from A, B, C the perpendiculars AF, BH, and CG. If the perpendicular from B falls at H between D and E, then will of the triangles so made BDH = AFD, and BHE = EGC (Theorems 6 and 15), whence follows that

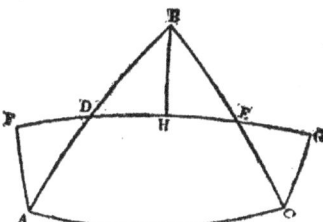

FIG. 16.

the surface of the triangle ABC equals that of the quadrilateral AFGC (Theorem 26).

If the point H coincides with the middle point E of the side BC (Fig.

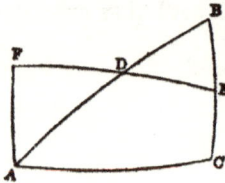

17), only two equal right-angled triangles, ADF and BDE, are made, by whose interchange the equivalence of the surfaces of the triangle ABC and the quadrilateral AFEC is established.

If, finally, the point H falls outside the triangle ABC (Fig. 18), the perpendicular CG goes, in consequence, through the triangle, and so we go

FIG. 17.

over from the triangle ABC to the quadrilateral AFGC by adding the

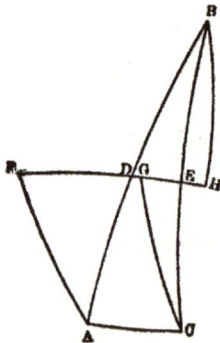

FIG. 18.

triangle FAD = DBH, and then taking away the triangle CGE = EBH.

Supposing in the spherical quadrilateral AFGC a great circle passed through the points A and G, as also through F and C, then will their arcs between AG and FC equal one another (Theorem 15), consequently also the triangles FAC and ACG be congruent (Theorem 15), and the angle FAC equal the angle ACG.

Hence follows, that in all the preceding cases, the sum of all three angles of the spherical triangle equals the sum of the two equal angles in the quadrilateral which are not the right angles.

Therefore we can, for every spherical triangle, in which the sum of the three angles is S, find a quadrilateral with equivalent surface, in which are two right angles and two equal perpendicular sides, and where the two other angles are each $\frac{1}{2}$S.

Let now ABCD (Fig. 19) be the spherical quadrilateral, where the sides AB = DC are perpendicular to BC, and the angles A and D each $\frac{1}{2}$S.

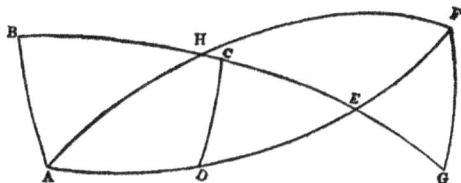

FIG. 19.

Prolong the sides AD and BC until they cut one another in E, and further beyond E, make DE = EF and let fall upon the prolongation of BC the perpendicular FG. Bisect the whole arc BG and join the mid-point H by great-circle-arcs with A and F.

The triangles EFG and DCE are congruent (Theorem 15), so FG = DC = AB.

The triangles ABH and HGF are likewise congruent, since they are right angled and have equal perpendicular sides, consequently AH and AF pertain to *one* circle, the arc AHF = π, ADEF likewise = π, the angle HAD = HFE = $\frac{1}{2}$S — BAH = $\frac{1}{2}$S — HFG = $\frac{1}{2}$S — HFE—EFG = $\frac{1}{2}$S—HAD—π+$\frac{1}{2}$S; consequently, angle HFE = $\frac{1}{2}$(S—π); or what is the same, this equals the size of the lune AHFDA, which again is equal to the quadrilateral ABCD, as we easily see if we pass over from the one to the other by first adding the triangle EFG and then BAH and thereupon taking away the triangles equal to them DCE and HFG.

Therefore $\frac{1}{2}$(S—π) is the size of the quadrilateral ABCD and at the same time also that of the spherical triangle in which the sum of the three angles is equal to S.

28. *If three planes cut each other in parallel lines, then the sum of the three surface angles equals two right angles.*

Let AA′, BB′ CC′ (Fig. 20) be three parallels made by the inter-section of planes (Theorem 25). Take upon them at random three

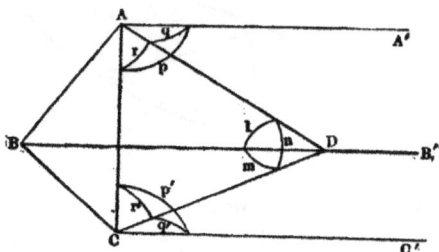

FIG. 20.

points A, B, C, and suppose through these a plane passed, which con-sequently will cut the planes of the parallels along the straight lines AB, AC, and BC. Further, pass through the line AC and any point D on the BB′, another plane, whose intersection with the two planes of the ·parallels AA′ and BB′, CC′ and BB′ produces the two lines AD and DC, and whose inclination to the third plane of the parallels AA′ and CC′ we will designate by w.

The angles between the three planes in which the parallels lie will be designated by X, Y, Z, respectively at the lines AA′, BB′, CC′; finally call the linear angles BDC $=$ a, ADC $=$ b, ADB $=$ c.

About A as center suppose a sphere described, upon which the inter-sections of the straight lines AC, AD AA′ with it determine a spherical triangle, with the sides p, q, and r. Call its size α. Opposite the side q lies the angle w, opposite r lies X, and consequently opposite p lies the angle $\pi + 2\alpha - w - $X, (Theorem 27).

In like manner CA, CD, CC′ cut a sphere about the center C, and determine a triangle of size β, with the sides p′, q′, r′, and the angles, w opposite q′, Z opposite r′, and consequently $\pi + 2\beta - w - $Z opposite p′.

Finally is determined by the intersection of a sphere about D with the lines DA, DB, DC, a spherical triangle, whose sides are l, m, n, and the angles opposite them $w + Z - 2\beta$, $w + X - 2\alpha$, and Y. Consequently its size $\delta = \frac{1}{2}($X$+$Y$+$Z$-\pi) - \alpha - \beta + w$.

Decreasing w lessens also the size of the triangles α and β, so that $\alpha + \beta - w$ can be made smaller than any given number.

In the triangle δ can likewise the sides l and m be lessened even to vanishing (Theorem 21), consequently the triangle δ can be placed with one of its sides l or m upon a great circle of the sphere as often as you choose without thereby filling up the half of the sphere, hence δ vanishes together with w; whence follows that necessarily we must have

$$X+Y+Z = \pi$$

29. *In a rectilineal triangle, the perpendiculars erected at the mid-points of the sides either do not meet, or they all three cut each other in one point.*

Having pre-supposed in the triangle ABC (Fig. 21), that the two perpendiculars ED and DF, which are erected upon the sides AB and BC at their mid points E and F, intersect in the point D, then draw within the angles of the triangle the lines DA, DB, DC.

In the congruent triangles ADE and BDE (Theorem 10), we have $AD = BD$, thus follows also that $BD = CD$; the triangle ADC is hence isosceles, consequently the perpendicular dropped from the vertex D upon the base AC falls upon G the mid-point of the base.

The proof remains unchanged also in the case when the intersection point D of the two perpendiculars ED and FD falls in the line AC itself, or falls without the triangle.

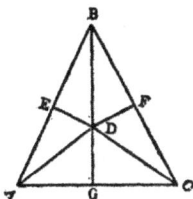

FIG. 21.

In case we therefore pre-suppose that two of those perpendiculars do not intersect, then also the third can not meet with them.

30. *The perpendiculars which are erected upon the sides of a rectilineal triangle at their mid-points, must all three be parallal to each other, so soon as the parallelism of two of them is pre-supposed.*

In the triangle ABC (Fig. 22) let the lines DE, FG, HK, be erected perpendicular upon the sides at their mid-points D, F, H. We will in the first place assume that the two perpendiculars DE and FG are parallel, cutting the line AB in L and M, and that the perpendicular HK lies between them. Within the angle BLE draw from the point L, at random, a straight line LG, which must cut FG somewhere in G, how small soever the angle of deviation GLE may be. (Theorem 16).

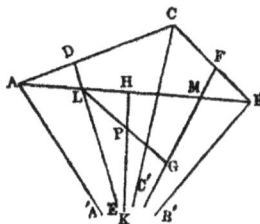

FIG. 22.

Since in the triangle LGM the perpendicular HK can not meet with MG (Theorem 29), therefore it must cut LG somewhere in P, whence follows, that HK is parallel to DE (Theorem 16), and to MG (Theorems 18 and 25).

Put the side $BC = 2a$, $AC = 2b$, $AB = 2c$, and designate the angles opposite these sides by A, B, C, then we have in the case just considered

$$A = \Pi(b) - \Pi(c),$$
$$B = \Pi(a) - \Pi(c),$$
$$C = \Pi(a) + \Pi(b),$$

as one may easily show with help of the lines AA′, BB′, CC′, which are drawn from the points A, B, C, parallel to the perpendicular HK and consequently to both the other perpendiculars DE and FG (Theorems 23 and 25).

Let now the two perpendiculars HK and FG be parallel, then can the third DE not cut them (Theorem 29), hence is it either parallel to them, or it cuts AA′.

The last assumption is not other than that the angle

$$C > \Pi(a) + \Pi(b.)$$

If we lessen this angle, so that it becomes equal to $\Pi(a) + \Pi(b)$, while we in that way give the line AC the new position CQ, (Fig. 23), and designate the size of the third side BQ by $2c'$, then must the angle CBQ at the point B, which is increased, in accordance with what is proved above, be equal to

$$\Pi(a) - \Pi(c') > \Pi(a) - \Pi(c),$$

whence follows $c' > c$ (Theorem 23).

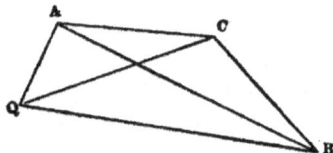

FIG. 23.

In the triangle ACQ are, however, the angles at A and Q equal, hence in the triangle ABQ must the angle at Q be greater than that at the point A, consequently is $AB > BQ$, (Theorem 9); that is $c > c'$.

31. *We call boundary line (oricycle) that curve lying in a plane for which all perpendiculars erected at the mid-points of chords are parallel to each other.*

In conformity with this definition we can represent the generation of a boundary line, if we draw to a given line AB (Fig. 24) from a given

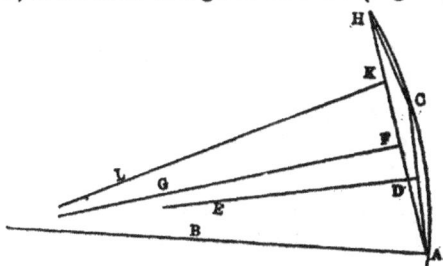

FIG. 24.

point A in it, making different angles CAB $= \Pi(a)$, chords AC $= 2a$; the end C of such a chord will lie on the boundary line, whose points we can thus gradually determine.

The perpendicular DE erected upon the chord AC at its mid-point D will be parallel to the line AB, which we will call the *Axis of the boundary line.* In like manner will also each perpendicular FG erected at the mid-point of any chord AH, be parallel to AB, consequently must this peculiarity also pertain to every perpendicular KL in general which is erected at the mid-point K of any chord CH, between whatever points C and H of the boundary line this may be drawn (Theorem 30). Such perpendiculars must therefore likewise, without distinction from AB, be called *Axes of the boundary line.*

32. *A circle with continually increasing radius merges into the boundary line.*

Given AB (Fig. 25) a chord of the boundary line; draw from the end-points A and B of the chord two axes AC and BF, which consequently will make with the chord two equal angles BAC $=$ ABF $= \alpha$ (Theorem 31).

Upon one of these axes AC, take anywhere the point E as center of a circle, and draw the arc AF from the initial point A of the axis AC to its intersection point F with the other axis BF.

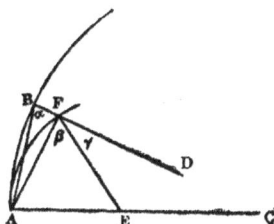

FIG. 25.

The radius of the circle, FE, corresponding to the point F will make on the one side with the chord AF an angle AFE $= \beta$, and on the other side with the axis BF, the angle EFD $= \gamma$. It follows that the angle between the two chords BAF $= a-\beta < \beta+\gamma-a$ (Theorem 22); whence follows, $a-\beta < \frac{1}{2}\gamma$.

Since now however the angle γ approaches the limit zero, as well in consequence of a moving of the center E in the direction AC, when F remains unchanged, (Theorem 21), as also in consequence of an approach of F to B on the axis BF, when the center E remains in its position (Theorem 22), so it follows, that with such a lessening of the angle γ, also the angle $a-\beta$, or the mutual inclination of the two chords AB and AF, and hence also the distance of the point B on the boundary line from the point F on the circle, tends to vanish.

Consequently one may also call the boundary-line *a circle with infinitely great radius*.

33. Let AA$'$ $=$ BB$'$ $= x$ (Figure 26), be two lines parallel toward the side from A to A$'$, which parallels serve as axes for the two boundary arcs (arcs on two boundary lines) AB$=s$, A$'$B$'=s'$, then is

$$s' = s e - x$$

Fig. 26.

where e is independent of the arcs s, s' and of the straight line x, the distance of the arc s' from s.

In order to prove this, assume that the ratio of the arc s to s' is equal to the ratio of the two whole numbers n and m.

Between the two axes AA$'$, BB$'$ draw yet a third axis CC$'$, which so cuts off from the arc AB a part AC $= t$ and from the arc A$'$B$'$ on the same side, a part A$'$C$' = t'$. Assume the ratio of t to s equal to that of the whole numbers p and q, so that·

$$s = \frac{n}{m}\,s', \quad t = \frac{p}{q}\,s.$$

Divide now s by axes into nq equal parts, then will there be mq such parts on s' and np on t.

However there correspond to these equal parts on s and t likewise equal parts on s' and t', consequently we have

$$\frac{t'}{t} = \frac{s'}{s}$$

Hence also wherever the two arcs t and t' may be taken between the two axes AA$'$ and BB$'$, the ratio of t to t' remains always the same, as

long as the distance x between them remains the same. If we therefore for $x = 1$, put $s = \mathrm{e}s'$, then we must have for every x

$$s' = s\mathrm{e}^{-x}.$$

Since e is an unknown number only subjected to the condition $\mathrm{e} > 1$, and further the linear unit for x may be taken at will, therefore we may, for the simplification of reckoning, so choose it that by e is to be understood the base of Napierian logarithms.

We may here remark, that $s' = 0$ for $x = \infty$, hence not only does the distance between two parallels decrease (Theorem 24), but with the prolongation of the parallels toward the side of the parallelism this at last wholly vanishes. Parallel lines have therefore the character of asymptotes.

34. *Boundary surface* (*orisphere*) we call that surface which arises from the revolution of the boundary line about one of its axes, which, together with all other axes of the boundary-line, will be also an axis of the boundary-surface.

A chord is inclined at equal angles to such axes drawn through its end-points, wheresoever these two end-points may be taken on the boundary-surface.

Let A, B, C, (Fig. 27), be three points on the boundary-surface;

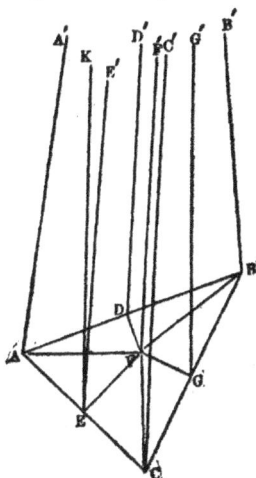

FIG. 27.

AA$'$, the axis of revolution, BB$'$ and CC$'$ two other axes, hence **AB** and AC chords to which the axes are inclined at equal angles A$'$AB $=$ B$'$BA, A$'$AC $=$ C$'$CA (Theorem 31.)

Two axes BB′, CC′, drawn through the end-points of the third chord BC, are likewise parallel and lie in one plane, (Theorem 25).

A perpendicular DD′ erected at the mid-point D of the chord AB and in the plane of the two parallels AA′, BB′, must be parallel to the three axes AA′, BB′, CC′, (Theorems 23 and 25); just such a perpendicular EE′ upon the chord AC in the plane of the parallels AA′, CC′ will be parallel to the three axes AA′, BB′, CC′, and the perpendicular DD′. Let now the angle between the plane in which the parallels AA′ and BB′ lie, and the plane of the triangle ABC be designated by $\Pi(a)$, where a may be positive, negative or null. If a is positive, then erect FD $= a$ within the triangle ABC, and in its plane, perpendicular upon the chord AB at its mid-point D.

Were a a negative number, then must FD $= a$ be drawn outside the triangle on the other side of the chord AB; when $a = 0$, the point F coincides with D.

In all cases arise two congruent right-angled triangles AFD and DFB, consequently we have FA $=$ FB.

Erect now at F the line FF′ perpendicular to the plane of the triangle ABC.

Since the angle D′DF $= \Pi(a)$, and DF $= a$, so FF′ is parallel to DD′ and the line EE′, with which also it lies in one plane perpendicular to the plane of the triangle ABC.

Suppose now in the plane of the parallels EE′, FF′ upon EF the perpendicular EK erected, then will this be also at right angles to the plane of the triangle ABC, (Theorem 13), and to the line AE lying in this plane, (Theorem 11); and consequently must AE, which is perpendicular to EK and EE′, be also at the same time perpendicular to FE, (Theorem 11). The triangles AEF and FEC are congruent, since they are right-angled and have the sides about the right angles equal, hence is

$$AF = FC = FB.$$

A perpendicular from the vertex F of the isosceles triangle BFC let fall upon the base BC, goes through its mid-point G; a plane passed through this perpendicular FG and the line FF′ must be perpendicular to the plane of the triangle ABC, and cuts the plane of the parallels BB′, CC′, along the line GG′, which is likewise parallel to BB′ and CC′, (Theorem 25); since now CG is at right angles to FG, and hence at the same time also to GG′, so consequently is the angle C′CG $=$ B′BG, (Theorem 23).

Hence follows, that for the boundary-surface each of the axes may be considered as axis of revolution.

Principal-plane we will call each plane passed through an axis of the boundary surface.

Accordingly every *Principal-plane* cuts the boundary-surface in the boundary line, while for another position of the cutting plane this in-tersection is a circle.

Three principal planes which mutually cut each other, make with each other angles whose sum is π, (Theorem 28).

These angles we will consider as angles in the boundary-triangle whose sides are arcs of the boundary-line, which are made on the bound-ary surface by the intersections with the three principal planes. Con-sequently the same interdependence of the angles and sides pertains to the boundary-triangles, that is proved in the ordinary geometry for the rectilineal triangle.

35. In what follows, we will designate the size of a line by a letter with an accent added, e. g. x', in order to indicate that this has a rela-tion to that of another line, which is represented by the same letter without accent x, which relation is given by the equation

$$\Pi(x) + \Pi(x') = \tfrac{1}{2}\pi.$$

Let now ABC (Fig. 28) be a rectilineal right-angled triangle, where the hypothenuse AB $=$ c, the other sides AC $=$ b, BC $=$ a, and the

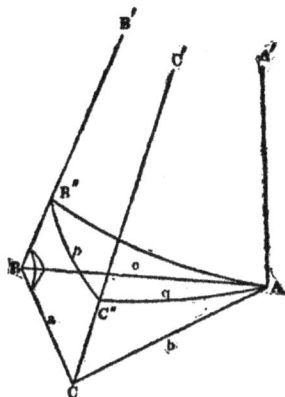

Fig. 28.

angles opposite them are

$$\text{BAC} = \Pi(a), \ \text{ABC} = \Pi(\beta).$$

At the point A erect the line AA′ at right angles to the plane of the triangle ABC, and from the points B and C draw BB′ and CC′ parallel to AA′.

The planes in which these three parallels lie make with each other the angles: $\Pi(a)$ at AA′, a right angle at CC′ (Theorems 11 and 13), consequently $\Pi(a')$ at BB′ (Theorem 28).

The intersections of the lines BA, BC, BB′ with a sphere described about the point B as center, determine a spherical triangle mnk, in which the sides are $mn = \Pi(c)$, $kn = \Pi(\beta)$, $mk = \Pi(a)$ and the opposite angles are $\Pi(b)$, $\Pi(a')$, $\frac{1}{2}\pi$.

Therefore we must, with the existence of a rectilineal triangle whose sides are a, b, c and the opposite angles $\Pi(a)$, $\Pi(\beta)$ $\frac{1}{2}\pi$, also admit the existence of a spherical triangle (Fig. 29) with the sides $\Pi(c)$, $\Pi(\beta)$, $\Pi(a)$ and the opposite angles $\Pi(b)$, $\Pi(a')$, $\frac{1}{2}\pi$.

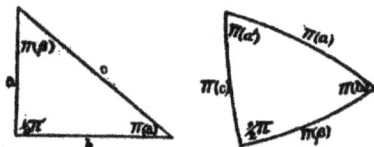

FIG. 29.

Of these two triangles, however, also inversely the existence of the spherical triangle necessitates anew that of a rectilineal, which in consequence, also can have the sides a, a′, β, and the oppsite angles $\Pi(b')$, $\Pi(c)$, $\frac{1}{2}\pi$.

Hence we may pass over from a, b, c, a, β, to b, a, c, β, a, and also to a, a′, β, b′, c.

Suppose through the point A (Fig. 28) with AA′ as axis, a boundary-surface passed, which cuts the two other axes BB′, CC′, in B″ and C″, and whose intersections with the planes the parallels form a boundary-triangle, whose sides are B″C″ $= p$, C″A $= q$, B″A $= r$, and the angles opposite them $\Pi(a)$, $\Pi(a')$, $\frac{1}{2}\pi$, and where consequently (Theorem 34):

$$p = r \sin \Pi(a), \quad q = r \cos \Pi(a).$$

Now break the connection of the three principal-planes along the line BB′, and turn them out from each other so that they with all the lines lying in them come to lie in one plane, where consequently the arcs p, q, r will unite to a single arc of a boundary-line, which goes through the point A and has AA′ for axis, in such a manner that (Fig. 30) on the one side will lie, the arcs q and p, the side b of the triangle, which is

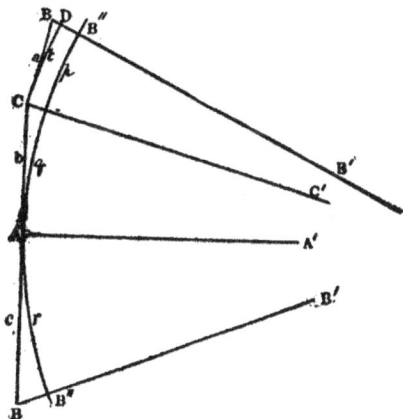

FIG. 30.

perpendicular to AA′ at A, the axis CC′ going from the end of b parallel to AA′ and through C″ the union point of p and q, the side a perpendicular to CC′ at the point C, and from the end-point of a the axis BB′ parallel to AA′ which goes through the end-point B″ of the arc p.

On the other side of AA′ will lie, the side c perpendicular to AA′ at the point A, and the axis BB′ parallel to AA′, and going through the end-point B″ of the arc r remote from the end point of b.

The size of the line CC″ depends upon b, which dependence we will express by CC″ $= f$ (b).

In like manner we will have BB′ $= f$ (c).

If we describe, taking CC′ as axis, a new boundary line from the point C to its intersection D with the axis BB′ and designate the arc CD by t, then is BD $= f$ (a).

$$BB′ = BD + DB″ = BD + CC″, \text{ consequently}$$
$$f(c) = f(a) + f(b).$$

Moreover, we perceive, that (Theorem 33)

$$t = pe^{f(b)} = r \sin \varPi(a)\, e^{f(b)}.$$

If the perpendicular to the plane of the triangle ABC (Fig. 28) were erected at B instead of at the point A, then would the lines c and r remain the same, the arcs q and t would change to t and q, the straight lines a and b into b and a, and the angle $\Pi(\alpha)$ into $\Pi(\beta)$, consequently we would have

$$q = r \sin \Pi(\beta) \; e^{f(a)},$$

whence follows by substituting the value of q,

$$\cos \Pi(\alpha) = \sin \Pi(\beta) \; e^{f(a)},$$

and if we change α and β into b' and c,

$$\sin \Pi(b) = \sin \Pi(c) \; e^{f(a)};$$

further, by multiplication with $e^{f(b)}$

$$\sin \Pi(b) \; e^{f(b)} = \sin \Pi(c) \; e^{f(c)}$$

Hence follows also

$$\sin \Pi(a) \; e^{f(a)} = \sin \Pi(b) \; e^{f(b)}.$$

Since now, however, the straight lines a and b are independent of one another, and moreover, for b=0, $f(b)=0$, $\Pi(b)=\frac{1}{2}\pi$, so we have for every straight line a

$$e^{-f(a)} = \sin \Pi(a).$$

Therefore,

$$\sin \Pi(c) = \sin \Pi(a) \sin \Pi(b),$$
$$\sin \Pi(\beta) = \cos \Pi(a) \sin \Pi(a).$$

Hence we obtain besides by mutation of the letters

$$\sin \Pi(a) = \cos \Pi(\beta) \sin \Pi(b),$$
$$\cos \Pi(b) = \cos \Pi(c) \cos \Pi(a),$$
$$\cos \Pi(a) = \cos \Pi(c) \cos \Pi(\beta).$$

If we designate in the right-angled spherical triangle (Fig. 29) the sides $\Pi(c)$, $\Pi(\beta)$, $\Pi(a)$, with the opposite angles $\Pi(b)$, $\Pi(a')$, by the letters a, b, c, A, B, then the obtained equations take on the form of those which we know as proved in spherical trigonometry for the right-angled triangle, namely,

$$\sin a = \sin c \sin A,$$
$$\sin b = \sin c \sin B,$$
$$\cos A = \cos a \sin B,$$
$$\cos B = \cos b, \sin A,$$
$$\cos c = \cos a, \cos b;$$

from which equations we can pass over to those for all spherical triangles in general.

Hence spherical trigonometry is not dependent upon whether in a rectilineal triangle the sum of the three angles is equal to two right angles or not.

86. We will now consider anew the right-angled rectilineal triangle ABC (Fig. 31), in which the sides are a, b, c, and the opposite angles $\Pi(a)$, $\Pi(\beta)$, $\frac{1}{2}\pi$.

Prolong the hypothenuse c through the point B, and make $BD=\beta$; at the point D erect upon BD the perpendicular DD', which consequently will be parallel to BB', the prolongation of the side a beyond the point B. Parallel to DD' from the point A draw AA', which is at the same time also parallel to CB', (Theorem 25), therefore is the angle

$$A'AD=\Pi(c+\beta),$$
$$A'AC=\Pi(b), \text{ consequently}$$
$$\Pi(b)=\Pi(a)+\Pi(c+\beta).$$

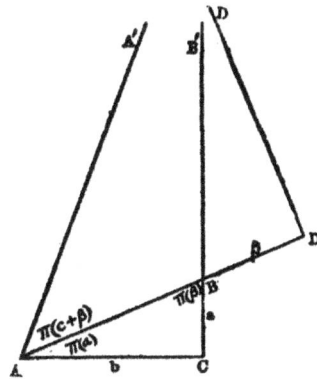
FIG. 31.

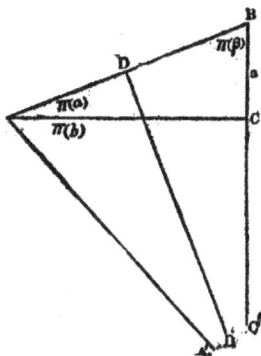
FIG. 32.

If from B we lay off β on the hypothenuse c, then at the end point D, (Fig. 32), within the triangle erect upon AB the perpendicular DD', and from the point A parallel to DD' draw AA', so will BC with its prolongation CC' be the third parallel; then is, angle $CAA'=\Pi$ (b), $DAA'=\Pi(c-\beta)$, consequently $\Pi(c-\beta)=\Pi(a)+\Pi(b)$. The last equation is then also still valid, when $c=\beta$, or $c<\beta$.

If $c=\beta$ (Fig. 33), then the perpendicular AA' erected upon AB at the point A is parallel to the side BC=a, with its prolongation, CC', consequently

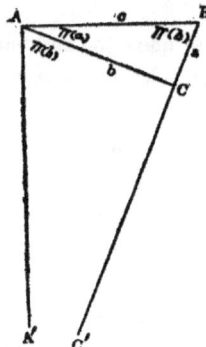

Fig. 33.

we have $\Pi(a)+\Pi(b)=\tfrac{1}{2}\pi$, whilst also $\Pi(c-\beta)=\tfrac{1}{2}\pi$, (Theorem 23).

If $c<\beta$, then the end of β falls beyond the point A at D (Fig. 34) upon the prolongation of the hypothenuse AB. Here the perpendicular DD' erected upon AD, and the line AA' parallel to it from A, will

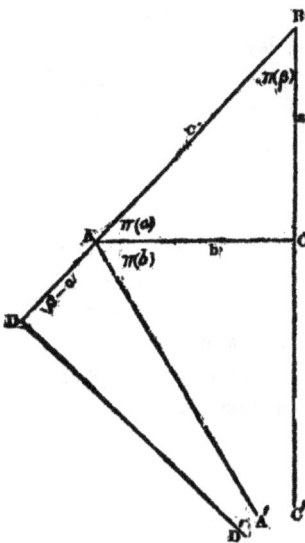

likewise be parallel to the side BC=a, with its prolongation CC'.

Here we have the angle DAA' $= \Pi$ (β—c), consequently

$$\Pi(a)+\Pi(b) =\pi-\Pi(\beta-c)=\Pi(c-\beta),$$

(Theorem 23).

The combination of the two equations found gives,

$$2\,\Pi(b)=\Pi(c-\beta)+\Pi(c+\beta),$$
$$2\,\Pi(a)=\Pi(c-\beta)-\Pi(c+\beta),$$

whence follows

$$\frac{\cos \Pi(b)}{\cos \Pi(a)} = \frac{\cos\left[\tfrac{1}{2}\Pi(c-\beta)+\tfrac{1}{2}\Pi(c+\beta)\right]}{\cos\left[\tfrac{1}{2}\Pi(c-\beta)-\tfrac{1}{2}\Pi(c+\beta)\right]}$$

Substituting here the value, (Theorem 35)

Fig. 34.

$$\frac{\cos \Pi(b)}{\cos \Pi(a)}=\cos\Pi(c),$$

we have $[\tan\tfrac{1}{2}\Pi(c)]^2=\tan\tfrac{1}{2}\Pi(c-\beta)\,\tan\tfrac{1}{2}\Pi(c+\beta).$

Since here β is an arbitrary number, as the angle $\Pi(\beta)$ at the one side of c may be chosen at will between the limits 0 and $\frac{1}{2}\pi$, consequently β between the limits 0 and ∞, so we may deduce by taking consecutively $\beta = c$, 2c, 3c, &c., that for every positive number n, $[\tan\frac{1}{2}\Pi(c)]^n = \tan\frac{1}{2}\Pi(nc)$.

If we consider n as the ratio of two lines x and c, and assume that $\cot\frac{1}{2}\Pi(c) = e^c$,

then we find for every line x in general, whether it be positive or negative, $\tan\frac{1}{2}\Pi(x) = e^{-x}$

where e may be any arbitrary number, which is greater than unity, since $\Pi(x) = 0$ for $x = \infty$.

Since the unit by which the lines are measured is arbitrary, so we may also understand by e the base of the Napierian Logarithms.

37. Of the equations found above in Theorem 35 it is sufficient to know the two following,

$$\sin \Pi(c) = \sin \Pi(a) \sin \Pi(b)$$
$$\sin \Pi(a) = \sin \Pi(b) \cos \Pi(\beta),$$

applying the latter to both the sides a and b about the right angle, in order from the combination to deduce the remaining two of Theorem 35, without ambiguity of the algebraic sign, since here all angles are acute.

In a similar manner we attain the two equations

(1.) $\tan \Pi(c) = \sin \Pi(a) \tan \Pi(a),$

(2.) $\cos \Pi(a) = \cos \Pi(c) \cos \Pi(\beta).$

We will now consider a rectilineal triangle whose sides are a, b, c, (Fig. 35) and the opposite angles A, B, C.

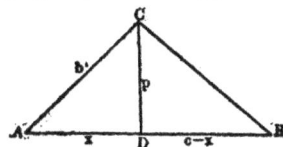

Fig. 35.

If A and B are acute angles, then the perpendicular p from the vertex of the angle C falls within the triangle and cuts the side c into two parts, x on the side of the angle A and c—x on the side of the angle B. Thus arise two right-angled triangles, for which we obtain, by application of equation (1),

$$\tan \Pi(a) = \sin B \tan \Pi(p),$$
$$\tan \Pi(b) = \sin A \tan \Pi(p),$$

which equations remain unchanged also when one of the angles, *e. g.* B, is a right angle (Fig. 36) or and obtuse angle (Fig. 37).

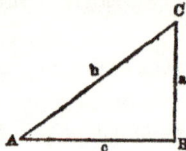

FIG. 36.

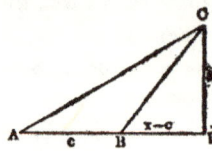

FIG. 37.

Therefore we have universally for every triangle

(3.) $\sin A \tan \Pi(a) = \sin B \tan \Pi(b)$.

For a triangle with acute angles A, B, (Fig. 35) we have also (Equation 2),

$$\cos \Pi(x) = \cos A \cos \Pi(b),$$
$$\cos \Pi(c-x) = \cos B \cos \Pi(a)$$

which equations also relate to triangles, in which one of the angles A or B is a right angle or an obtuse angle.

As example, for $B = \frac{1}{2}\pi$ (Fig. 36) we must take $x = c$, the first equation then goes over into that which we have found above as Equation 2, the other, however, is self-sufficing.

For $B > \frac{1}{2}\pi$ (Fig. 37) the first equation remains unchanged, instead of the second, however, we must write correspondingly

$$\cos \Pi(x-c) = \cos(\pi - B) \cos \Pi(a);$$

but we have $\cos \Pi(x-c) = -\cos \Pi(c-x)$

(Theorem 23), and also $\cos(\pi - B) = -\cos B$.

If A is a right or an obtuse angle, then must $c - x$ and x be put for x and $c - x$, in order to carry back this case upon the preceding.

In order to eliminate x from both equations, we notice that (Theorem 36)

$$\cos \Pi(c-x) = \frac{1 - [\tan \frac{1}{2}\Pi(c-x)]^2}{1 + [\tan \frac{1}{2}\Pi(c-x)]^2}$$
$$= \frac{1 - e^{2x-2c}}{1 + e^{2x-2c}}$$
$$= \frac{1 - [\tan \frac{1}{2}\Pi(c)]^2 [\cot \frac{1}{2}\Pi(x)]^2}{1 + [\tan \frac{1}{2}\Pi(c)]^2 [\cot \frac{1}{2}\Pi(x)]^2}$$
$$= \frac{\cos \Pi(c) - \cos \Pi(x)}{1 - \cos \Pi(c)\cos \Pi(x)}$$

If we substitute here the expression for $\cos \Pi(x)$, $\cos\Pi(c - x)$, we obtain

$$\cos\Pi(c) = \frac{\cos \Pi(a) \cos B + \cos\Pi(b) \cos A}{1 + \cos\Pi(a) \cos\Pi(b) \cos A \cos B}$$

whence follows

$$\cos \Pi(a) \cos B = \frac{\cos \Pi(c) - \cos A \cos \Pi(b)}{1 - \cos A \cos \Pi(b) \cos \Pi(c)}$$

and finally

$$[\sin \Pi(c)]^2 = [1 - \cos B \cos \Pi(c) \cos \Pi(a)][1 - \cos A \cos \Pi(b) \cos \Pi(c)]$$

In the same way we must also have

(4.)

$$[\sin \Pi(a)]^2 = [1 - \cos C \cos \Pi(a) \cos \Pi(b)][1 - \cos B \cos \Pi(c) \cos \Pi(a)]$$
$$[\sin \Pi(b)]^2 = [1 - \cos A \cos \Pi(b) \cos \Pi(c)][1 - \cos C \cos \Pi(a) \cos \Pi(b)]$$

From these three equations we find

$$\frac{[\sin \Pi(b)]^2 [\sin\Pi(c)]^2}{[\sin\Pi(a)]^2} = [1 - \cos A \cos \Pi(b) \cos \Pi(c)]^2.$$

Hence follows without ambiguity of sign,

(5.) $$\cos A \cos \Pi(b) \cos \Pi(c) + \frac{\sin \Pi(b) \sin \Pi(c)}{\sin \Pi(a)} = 1.$$

If we substitute here the value of $\sin \Pi(c)$ corresponding to equation (3.)

$$\sin \Pi(c) = \frac{\sin A}{\sin C} \tan \Pi(a) \cos \Pi(c)$$

then we obtain

$$\cos \Pi(c) = \frac{\cos \Pi(a) \sin C}{\sin A \sin \Pi(b) + \cos A \sin C \cos\Pi(a) \cos \Pi(b)};$$

but by substituting this expression for $\cos \Pi(c)$ in equation (4),

(6.) $$\cot A \sin C \sin \Pi(b) + \cos C = \frac{\cos \Pi(b)}{\cos \Pi(a)}$$

By elimination of $\sin \Pi(b)$ with help of the equation (3) comes

$$\frac{\cos \Pi(a)}{\cos \Pi(b)} \cos C = 1 - \frac{\cos A}{\sin B} \sin C \sin \Pi(a).$$

In the meantime the equation (6) gives by changing the letters,

$$\frac{\cos \Pi(a)}{\cos \Pi(b)} = \cot B \sin C \sin \Pi(a) + \cos C.$$

From the last two equations follows,

(7.) $\cos A + \cos B \cos C = \dfrac{\sin B \sin C}{\sin \Pi(a)}$

All four equations for the interdependence of the sides a, b, c, and the opposite angles A, B, C, in the rectilineal triangle will therefore be, [Equations (3), (5), (6), (7).]

(8.)
$$\begin{cases} \sin A \tan \Pi(a) = \sin B \tan \Pi(b), \\[2mm] \cos A \cos \Pi(b) \cos \Pi(c) + \dfrac{\sin \Pi(b) \sin \Pi(c)}{\sin \Pi(a)} = 1, \\[2mm] \cot A \sin C \sin \Pi(b) + \cos C = \dfrac{\cos \Pi(b)}{\cos \Pi(a)}, \\[2mm] \cos A + \cos B \cos C = \dfrac{\sin B \sin C}{\sin \Pi(a)}. \end{cases}$$

If the sides a, b, c, of the triangle are very small, we may content our-selves with the approximate determinations.   (Theorem 36.)

$$\cot \Pi(a) = a,$$
$$\sin \Pi(a) = 1 - \tfrac{1}{2}a^2$$
$$\cos \Pi(a) = a,$$

and in like manner also for the other sides b and c.

The equations 8 pass over for such triangles into the following:

$$b \sin A = a \sin B,$$
$$a^2 = b^2 + c^2 - 2bc \cos A,$$
$$a \sin (A + C) = b \sin A,$$
$$\cos A + \cos (B + C) = 0.$$

Of these equations the first two are assumed in the ordinary geom-etry; the last two lead, with the help of the first, to the conclusion

$$A + B + C = \pi.$$

Therefore the imaginary geometry passes over into the ordinary, when we suppose that the sides of a rectilineal triangle are very small.

I have, in the scientific bulletins of the University of Kasan, pub-lished certain researches in regard to the measurement of curved lines, of plane figures, of the surfaces and the volumes of solids, as well as in relation to the application of imaginary geometry to analysis.

The equations (8) attain for themselves already a sufficient foundation for considering the assumption of imaginary geometry as possible. Hence there is no means, other than astronomical observations, to use for judging of the exactitude which pertains to the calculations of the ordinary geometry.

This exactitude is very far-reaching, as I have shown in one of my investigations, so that, for example, in triangles whose sides are attain-able for our measurement, the sum of the three angles is not indeed dif-ferent from two right-angles by the hundredth part of a second.

In addition, it is worthy of notice that the four equations (8) of plane geometry pass over into the equations for spherical triangles, if we put $a \sqrt{-1}$, $b \sqrt{-1}$, $c \sqrt{-1}$, instead of the sides a, b, c; with this change, however, we must also put

$$\sin \varPi (a) = \frac{1}{\cos (a),}$$

$$\cos \varPi (a) = (\sqrt{-1}) \tan a,$$

$$\tan \varPi (a) = \frac{1}{\sin a \, (\sqrt{-1}),}$$

and similarly also for the sides b and c.

In this manner we pass over from equations (8) to the following:

$$\sin A \sin b = \sin B \sin a,$$
$$\cos a = \cos b \cos c + \sin b \sin c \cos A,$$
$$\cot A \sin C + \cos C \cos b = \sin b \cot a,$$
$$\cos A = \cos a \sin B \sin C - \cos B \cos C.$$

TRANSLATOR'S APPENDIX.

ELLIPTIC GEOMETRY.

Gauss himself never published aught upon this fascinating subject, Geometry Non-Euclidean; but when the most extraordinary pupil of his long teaching life came to read his inaugural dissertation before the Philosophical Faculty of the University of Goettingen, from the three themes submitted it was the choice of Gauss which fixed upon the one "Ueber die Hypothesen welche der Geometrie zu Grunde liegen."

Gauss was then recognized as the most powerful mathematician in the world. I wonder if he saw that here his pupil was already beyond him, when in his sixth sentence Riemann says, "therefore space is only a special case of a three-fold extensive magnitude," and continues: "From this, however, it follows of necessity, that the propositions of geometry can not be deduced from general magnitude-ideas, but that those peculiarities through which space distinguishes itself from other thinkable threefold extended magnitudes can only be gotten from experience. Hence arises the problem, to find the simplest facts from which the metrical relations of space are determinable — a problem which from the nature of the thing is not fully determinate; for there may be obtained several systems of simple facts which suffice to determine the metrics of space; that of Euclid as weightiest is for the present aim made fundamental. These facts are, as all facts, not necessary, but only of empirical certainty; they are hypotheses. Therefore one can investigate their probability, which, within the limits of observation, of course is very great, and after this judge of the allowability of their extension beyond the bounds of observation, as well on the side of the immeasurably great as on the side of the immeasurably small."

Riemann extends the idea of curvature to spaces of three and more dimensions. The curvature of the sphere is constant and positive, and on it figures can freely move without deformation. The curvature of the plane is constant and zero, and on it figures slide without stretching. The curvature of the two-dimensional space of Lobachevski and

Bolyai completes the group, being constant and negative, and in it figures can move without stretching or squeezing.   As thus corresponding
to the sphere it is called the pseudo-sphere.

In the space in which we live, we suppose we can move without deformation.   It would then, according to Riemann, be a special case of
a space of constant curvature.   We presume its curvature null.   At
once the supposed fact that our space does not interfere to squeeze us
or stretch us when we move, is envisaged as a peculiar property of our
space.   But is it not absurd to speak of space as interfering with any-
thing?   If you think so, take a knife and a raw potato, and try to cut
it into a seven-edged solid.

Further on in this astonishing discourse comes the epoch-making idea,
that though space be unbounded, it is not therefore infinitely great.
Riemann says: "In the extension of space-constructions to the immeasurably great, the unbounded is to be distinguished from the infinite; the first pertains to the relations of extension, the latter to the
size-relations.

"That our space is an unbounded three-fold extensive manifoldness, is
a hypothesis, which is applied in each apprehension of the outer world,
according to which, in each moment, the domain of actual perception is
filled out, and the possible places of a sought object constructed, and
which in these applications is continually confirmed.   The unbounded-
ness of space possesses therefore a greater empirical certainty than any
outer experience.   From this however the Infinity in no way follows.
Rather would space, if one presumes bodies independent of place, that
is ascribes to it a constant curvature, necessarily be finite so soon as this
curvature had ever so small a positive value.   One would, by extend-
ing the beginnings of the geodesics lying in a surface-element, obtain
an unbounded surface with constant positive curvature, therefore a surface which in a homaloidal three-fold extensive manifoldness would
take the form of a sphere, and so is finite."

Here we have for the first time in human thought the marvelous perception that universal space may yet be only finite.

Assume that a straight line is uniquely determined by two points, but
take the contradictory of the axiom that a straight line is of infinite
size; then the straight line returns into itself, and two having inter-
sected get back to that intersection point.

BIBLIOGRAPHY.

A bibliography of non-Euclidean literature down to the year 1878 was given by Halsted, *"American Journal of Mathematics,"* vols. i, ii, containing 81 authors and 174 titles, and reprinted in the collected works of Lobachevski (Kazan, 1886) giving 124 authors and 272 titles. This was incorporated in Bonola's Bibliography of the Foundations of Geometry (1899) reprinted (1902) at Kolozsvár in the Bolyai Memorial Volume. In 1911 appeared the volume: Bibliography of Non-Euclidean Geometry by Duncan M. Y. Sommerville; London, Harrison and Sons.

The Introduction says: "The present work was begun about nine years ago. It was intended as a continuation of Halsted's bibliography, but it soon became evident that the growth of the subject rendered such diffuse treatment practically impossible, and short abstracts of the works would have to be dispensed with. The object is to produce as far as possible a complete repository of the titles of all works from the earliest times up to the present which deal with the extended conception of space, and to form a guide to the literature in an easily accessible form. It includes the theory of parallels, non-euclidean geometry, the foundations of geometry, and space of *n* dimensions."

In 1913 Teubner issued in two parts Paul Stäckel's important book: Wolfgang und Johann Bolyai. Geometrische Untersuchungen. John compares Lobachevski's researches with his own. The profound philosophic import of non-euclidean geometry forms an integrant part of "The Foundations of Science," by H. Poincaré; Vol. I of the series Science and Education, The Science Press, New York City, 1914. The Transactions of the Royal Society of Canada, Vol. XII, Section III, contains a striking Presidential Address by Alfred Baker on The Foundations of Geometry. Of the cognate works issued by The Open Court Pub. Co., we mention only Euclid's Parallel Postulate by Withers. Scores of errors are pointed out in "Non-Euclidean Geometry in the Encyclopædia Britannica," Science, May 10, 1912.

And now at last the theory of relativity has made non-euclidean geometry a powerful machine for advance in physics.

Says Vladimir Varićak in a remarkable lecture, "Ueber die nichteuklidische Interpretation der Relativtheorie," (Jahresber. D. Math. Ver., 21, 103-127),

I postulated that the phenomena happened in a Lobachevski space, and reached by very simple geometric deduction the formulas of the relativity theory. Assuming non-euclidean terminology, the formulas of the relativity theory become not only essentially simplified, but capable of a geometric interpretation wholly analogous to the interpretation of the classic theory· in the euclidean geometry. And this analogy often goes so far, that the very wording of the theorems of the classic theory may be left unchanged.

www.ingramcontent.com/pod-product-compliance
Lightning Source LLC
Chambersburg PA
CBHW022107210326
41520CB00045B/541